FIRST SUPPLEMENT
TO THE
THIRD EDITION

FOOD
CHEMICALS
CODEX

COMMITTEE ON FOOD CHEMICALS CODEX

Food and Nutrition Board
Commission on Life Sciences
National Research Council

NATIONAL ACADEMY PRESS
Washington, D.C.   1983

NATIONAL ACADEMY PRESS   **2101 CONSTITUTION AVENUE, NW**   **WASHINGTON, DC 20418**

NOTICE The project that is the subject of this report was approved by the Governing Board of the National Research Council, whose members are drawn from the Councils of the National Academy of Sciences, the National Academy of Engineering, and the Institute of Medicine. The members of the Committee responsible for the report were chosen for their special competences and with regard for appropriate balance.

NATIONAL RESEARCH COUNCIL The National Research Council was established by the National Academy of Sciences in 1916 to associate the broad community of science and technology with the Academy's purposes of furthering knowledge and of advising the federal government. The Council operates in accordance with general policies determined by the Academy under the authority of its congressional charter of 1863, which establishes the Academy as a private, nonprofit, self-governing membership corporation. The Council has become the principal operating agency of both the National Academy of Sciences and the National Academy of Engineering in the conduct of their services to the government, the public, and the scientific and engineering communities. It is administered jointly by both Academies and the Institute of Medicine. The National Academy of Engineering and the Institute of Medicine were established in 1964 and 1970, respectively, under the charter of the National Academy of Sciences.

FOOD AND NUTRITION BOARD The Food and Nutrition Board was established in 1940. It is a division of the Commission on Life Sciences of the National Research Council.

The Board serves as an advisory body in the field of food and nutrition. It promotes needed research and helps interpret nutritional science in the interests of public welfare. The Board acts in response to requests from public agencies and, at times, on its own initiative.

The Board is active in areas of dietary guidelines, nutrition and health, food safety, food chemicals specifications, food resources, and international nutrition programs. It has established, among other important guides, recommended dietary allowances, principles and procedures for the evaluation of the safety of foods, specifications of identity and purity for food chemicals, guidelines for nutrient fortification of foods, and recommendations for maternal and infant nutrition. The Food and Nutrition Board draws upon the knowledge and expertise available from the combined resources of academia, government, and industry.

Financial support for the work of the Board is primarily provided by government contracts and grants. In addition, uncommitted support is provided by private foundations and industrial organizations.

Through members of its liaison panels, technical input in aspects of nutrition, food safety, food technology, and food processing is provided.

This study is supported by U.S. Food and Drug Administration Contract No. 223-78-2053 (formerly Grant No. FD 00213).

COMPLIANCE WITH FEDERAL STATUTES The fact that an article appears in the *Food Chemicals Codex* or its supplements does not exempt it from compliance with requirements of acts of Congress, with regulations and rulings issued by agencies of the United States Government under authority of these acts, or with requirements and regulations of governments in other countries that have adopted the *Food Chemicals Codex*. Revisions of the federal requirements that affect the Codex specifications will be included in Codex supplements as promptly as practicable.

LIBRARY OF CONGRESS CATALOG CARD NUMBER 81-38403
INTERNATIONAL STANDARD BOOK NUMBER 0-309-03090-0

National Academy Press

 The National Academy Press was created by the National Academy of Sciences to publish the reports issued by the Academy and by the National Academy of Engineering, the Institute of Medicine, and the National Research Council, all operating under the charter granted to the National Academy of Sciences by the Congress of the United States.

Contents

To the memory of
JUSTIN L. POWERS, Ph.D.
1895-1981
First Director of the *Food Chemicals Codex,* 1961-66

Organization of the *Food Chemicals Codex* (1981–83)

COMMITTEE ON FOOD CHEMICALS CODEX

James R. Kirk
  Chairman
Samuel Tuthill
  Vice-Chairman
Frank L. Boyd
Bruce H. Campbell

Jack P. Fletcher
Sol W. Gunner
Thomas Medwick
Fred A. Morecombe
Jessie McGowan Norris
Jane C. Sheridan
Jan Stofberg

Robert A. Mathews, *Staff Officer*
Betty C. Guyot, *Secretary*

FOOD AND NUTRITION BOARD

Irwin H. Rosenberg
  Chairman (1982–83)

Alfred E. Harper
  Chairman (1981–82)

Myrtle L. Brown, *Executive Secretary*
Sushma Palmer, *Executive Director* (from July 1983)

COMMISSION ON LIFE SCIENCES

Frederick C. Robbins
  Chairman

Alvin G. Lazen, *Executive Director*

Additions, Changes, and Corrections

Additions, changes, and corrections listed herein constitute revisions in the *Food Chemicals Codex*, Third Edition (FCC III). Page numbers refer to FCC III unless indicated by a reference to pages in THIS SUPPLEMENT.

1/ General Provisions Applying to Specifications, Tests, and Assays of the Food Chemicals Codex

No Change.

2/ Monographs

Insert the following new monograph to precede the monograph entitled *Adipic Acid*, page 11:

Acid Hydrolyzed Proteins

Hydrolyzed Vegetable Protein (HVP); Hydrolyzed Plant Protein (HPP); Hydrolyzed Milk Protein

DESCRIPTION

Acid hydrolyzed proteins are composed primarily of amino acids and salts resulting from the acid-catalyzed breakdown of peptide bonds present in edible proteinaceous materials. In processing, the protein hydrolysates may be treated with safe and suitable alkaline materials. The edible proteinaceous materials used as raw materials are derived from corn, soy, wheat, yeast, peanut, rice, or other safe and suitable vegetable sources, or from milk. Individual products may be in liquid, paste, powder, or granular form. The pH of a 2% solution in water is between 4.3 and 7.0.

REQUIREMENTS

Calculate all analyses on the dried basis. Liquid and paste samples should be evaporated to dryness on a steam bath, then, as for the powdered and granular forms, dried to constant weight at 105° (see *General Provisions Applying to Specifications, Tests, and Assays of the Food Chemicals Codex*, page 1).

Assay (Total Nitrogen)   Not less than 3.25% total nitrogen.
α-Amino Nitrogen   Not less than 2.0%.
Arsenic   Not more than 3 ppm.

Aspartic Acid   Not more than 6.0% as $C_4H_7NO_4$ and not more than 15.0% of the total amino acids.
Glutamic Acid   Not more than 20.0% as $C_5H_9NO_4$ and not more than 35.0% of the total amino acids.
Heavy Metals (as Pb)   Not more than 0.002%.
Insoluble Matter   Not more than 1%.
Lead   Not more than 10 ppm.
Sodium   Not more than 25.0%.

TESTS

Assay (Total Nitrogen)   Proceed as directed under *Nitrogen Determination*, page 521.

α-Amino Nitrogen   Transfer 7 to 25 g, accurately weighed, into a 500-ml volumetric flask with the aid of several 50-ml portions of warm ammonia-free water, dilute to volume with water, and mix. Neutralize 20.0 ml of the solution with 0.2 *N* barium hydroxide or 0.2 *N* sodium hydroxide, using phenolphthalein TS as indicator, and add 10 ml of freshly prepared phenolphthalein-formol solution (50 ml of 40% formic acid containing 1 ml of 0.05% phenolphthalein in 50% alcohol neutralized exactly to pH 7 with 0.2 *N* barium hydroxide or sodium hydroxide). Titrate with 0.2 *N* barium hydroxide to a distinct red color, add a small but accurately measured volume of 0.2 *N* barium hydroxide in excess, and back titrate to neutrality with 0.2 *N* hydrochloric acid. Conduct a blank titration using the same reagents, with 20 ml of water in place of the test solution. Each ml of 0.2 *N* barium hydroxide is equivalent to 2.8 mg of α-amino nitrogen.

Arsenic   A *Sample Solution* prepared as directed for organic compounds meets the requirements of the *Arsenic Test*, page 464.

3

Aspartic and Glutamic Acids

Apparatus Use an ion-exchange amino acid analyzer, equipped with sulfonated polystyrene columns, in which the effluent from the sample is mixed with ninhydrin reagent and the absorbance of the resultant color is measured continuously and automatically at 570 and 440 nm by a recording photometer.

Standard Solution Prepare a standard mixture of amino acids as follows: Weigh 1250 ± 2 μmol of each amino acid, and place in a 500-ml volumetric flask. Fill the flask half-full with water, and add 5 ml of concentrated hydrochloric acid to dissolve the less-soluble amino acids. Prepare the standard for analysis by diluting 1 ml of this solution with 4 ml of 0.2 N sodium citrate, pH 2.2, buffer. Each 2-ml aliquot contains 1.00 μmol of each amino acid.

Sample Preparation Accurately weigh 5 mg of the sample, and dilute to exactly 5 ml with 0.2 N sodium citrate, pH 2.2, buffer. Remove any insoluble material by centrifugation or filtration.

Procedure Using 2-ml aliquots of the *Standard Solution* and *Sample Preparation*, proceed as directed according to the apparatus manufacturer's instructions. From the chromatograms thus obtained, match the retention times produced by the *Standard Solution* with those produced by the *Sample Solution*, and identify the peaks produced by aspartic acid and glutamic acid. Record the area of the respective amino acid peak from the sample as A_A, and that from the standards as A_S.

Calculations Calculate the respective concentration, C_A, in μmol per ml, of aspartic acid and glutamic acid in the *Sample Preparation* by the formula $A_A \times C_S / A_S$, in which C_S is the concentration, in μmol per ml, of the respective amino acid in the *Standard Solution*.

Calculate the respective percentage of aspartic acid and glutamic acid, on the basis of total amino acids, by the formula $16 \times C_A / N_T$, in which N_T is the percentage of total nitrogen determined in the *Assay*.

Calculate the respective percentage of aspartic acid and glutamic acid in the sample by the formula $100 \times C_A / S_W$, in which S_W is the weight of the sample taken, in g.

Heavy Metals Prepare and test a 1-g sample as directed in *Method II* under the *Heavy Metals Test*, page 513, using 20 μg of lead ion (Pb) in the control (*Solution A*).

Insoluble Matter Transfer about 5 g, accurately weighed, into a 250-ml Erlenmeyer flask, add 75 ml of water, cover the flask with a watch glass, and boil gently for 2 min. Filter the solution through a tared filtering crucible, dry at 105° for 1 h, cool, and weigh.

Lead A *Sample Solution* prepared as directed for organic compounds meets the requirements of the *Lead Limit Test*, page 518, using 10 μg of lead ion (Pb) in the control.

Sodium

Spectrophotometer Use any suitable atomic absorption spectrophotometer.

Standard Solution Transfer 25.42 mg of reagent-grade sodium chloride, accurately weighed, into a 1000-ml volumetric flask, dissolve in and dilute to volume with deionized water, and mix. Transfer 5.0 ml of this solution to a second 1000-ml volumetric flask, dilute to volume with deionized water, and mix. Each ml contains 0.5 μg of Na.

Sample Solution Transfer 1.00 ± 0.05 g of previously dried sample, accurately weighed, into a silica or porcelain dish. Ash in a muffle furnace at 246°–260° for 2–4 h. Allow the ash to cool, and dissolve in 5 ml of 20% HCl, warming the solution if necessary to complete solution of the residue. Filter the solution through acid-washed filter paper into a 500-ml volumetric flask. Wash the filter paper with hot water, dilute to volume, and mix.

Procedure Determine the absorbance of each solution at 589.0 nm, following the manufacturer's instructions for optimum operation of the spectrophotometer. The absorbance produced by the *Sample Solution* does not exceed that of the *Standard Solution*.

Packaging and Storage Store in well-closed containers.
Functional Use in Foods Flavoring agent.

Aluminum Ammonium Sulfate, page 14

Replace the *Test* entitled *Assay*, page 15, with the following (note that the buffer is added prior to the boiling step):

Assay Weigh accurately about 1 g of sample, dissolve in 50 ml of water, add 50.0 ml of 0.05 M disodium EDTA and 20 ml of pH 4.5 buffer solution (77.1 g of ammonium acetate and 57 ml of glacial acetic acid in 1000 ml of solution), and boil gently for 5 min. Cool, and add 50 ml of alcohol and 2 ml of dithizone TS. Titrate with 0.05 M zinc sulfate to a bright rose-pink color, and perform a blank determination (see page 2). Each ml of 0.05 M disodium EDTA is equivalent to 22.67 mg of $AlNH_4(SO_4)_2 \cdot 12H_2O$.

Aluminum Potassium Sulfate, page 15

Replace the *Test* entitled *Assay* with the following (note that the buffer is added prior to the boiling step):

Assay Weigh accurately about 1 g of sample, dissolve in 50 ml of water, add 50.0 ml of 0.05 M disodium EDTA and 20 ml of pH 4.5 buffer solution (77.1 g of ammonium acetate and 57 ml of glacial acetic acid in 1000 ml of solution), and boil gently for 5 min. Cool, and add 50 ml of alcohol and 2 ml of dithizone TS. Titrate with 0.05 M zinc sulfate to a bright rose-pink color, and perform a blank determination (see page 2). Each ml of 0.05 M disodium EDTA is equivalent to 23.72 mg of $AlK(SO_4)_2 \cdot 12H_2O$.

Aluminum Sodium Sulfate, page 16

Replace the *Test* entitled *Assay* with the following (note that the buffer is added prior to the boiling step):

Assay Weigh accurately about 500 mg of sample previously dried as directed in the test for *Loss on Drying*, moisten with 1 ml of acetic acid, and dissolve in 50 ml of water, warming gently on a steam bath until solution is complete. Cool, neutralize with ammonia TS, add 50.0 ml of 0.05 M disodium EDTA and 20 ml of pH 4.5 buffer solution (77.1 g of ammonium acetate and 57 ml of glacial acetic acid in 1000 ml of solution), and boil gently for 5 min. Cool, and add 50 ml of alcohol and 2 ml of dithizone TS. Titrate with 0.05 M zinc sulfate to a bright rose-pink color, and perform a blank determination (see page 2). Each ml of 0.05 M disodium EDTA is equivalent to 12.10 mg of $AlNa(SO_4)_2$.

Aluminum Sulfate, page 17

Replace the *Test* entitled *Assay* with the following (note that the buffer is added prior to the boiling step):

Assay Weigh accurately an amount of sample equivalent to about 4 g of $Al_2(SO_4)_3$, transfer to a 250-ml volumetric flask, dissolve in water, dilute to volume, and mix. Pipet 10 ml of this solution into a 250-ml beaker, add 25.0 ml of 0.05 M disodium EDTA and 20 ml of pH 4.5 buffer solution (77.1 g of ammonium acetate and 57 ml of glacial acetic acid in 1000 ml of solution), and boil gently for 5 min. Cool, and add 50 ml of alcohol and 2 ml of dithizone TS. Titrate with 0.05 M zinc sulfate to a bright rose-pink color, and perform a blank determination (see page 2). Each ml of 0.05 M disodium EDTA is equivalent to 8.554 mg of $Al_2(SO_4)_3$ or to 16.66 mg of $Al_2(SO_4)_3 \cdot 18H_2O$.

Ammonium Bicarbonate, page 19

Replace the *Test* entitled *Assay* with the following:

Assay Weigh accurately about 3 g of sample, and dissolve it in 40 ml of water. Add 2 drops of methyl red TS, and titrate with 1 N hydrochloric acid. Add the acid slowly, with constant stirring, until the solution becomes faintly pink. Heat the solution to boiling, cool, and continue the titration until the faint pink color no longer fades after boiling. Each ml of 1 N hydrochloric acid is equivalent to 79.06 mg of NH_4HCO_3.

Aspartame, page 28

Change the *Description* to read:

DESCRIPTION

A white, odorless, crystalline powder having a sweet taste. It is sparingly soluble in water and slightly soluble in alcohol. The pH of a 0.8% aqueous solution is between 4.5 and 6.0.

Add to the *Requirement* entitled *Identification* a paragraph C to read:

C. Identify aspartame by comparing its infrared absorption spectrum with the spectrum shown on page 30, THIS SUPPLEMENT. The spectrum is obtained from a sample pelleted with KBr.

Change the *Requirement* entitled *Specific Rotation* to read:

Specific Rotation $[\alpha]_D^{20°}$ Between $+14.5°$ and $+16.5°$, calculated on the dried basis.

Insert the following new monograph to precede the monograph entitled *Azodicarbonamide*, page 31.

Autolyzed Yeast Extract

DESCRIPTION

Autolyzed yeast extracts are composed primarily of (a) amino acids, peptides, and salts resulting from the acid-catalyzed hydrolysis of polypeptide bonds in naturally occurring enzymes present in the edible yeast and (b) the water-soluble components of the yeast cell. Food-grade salt may be added during processing. Individual products may be in liquid, paste, powder, or granular form. The pH of a 2% solution in water is between 4.5 and 6.0.

REQUIREMENTS

Calculate all determinations on the dried basis. Liquid and paste samples should be evaporated to dryness on a steam bath, then, as for the powdered and granular forms, dried to constant weight at 105° (see *General Provisions Applying to Specifications, Tests, and Assays of the Food Chemicals Codex*, page 1).

Assay (Total Nitrogen) Not less than 9.0% total nitrogen.
α-Amino Nitrogen Not less than 3.5%.
Arsenic Not more than 3 ppm.

Aspartic Acid   Not more than 8.0% as $C_4H_7NO_4$ and not more than 12.0% of the total amino acids.
Glutamic Acid   Not more than 12.0% as $C_5H_9NO_4$ and not more than 20.0% of the total amino acids.
Heavy Metals (as Pb)   Not more than 0.002%.
Insoluble Matter   Not more than 1%.
Lead   Not more than 10 ppm.
Sodium   Not more than 20.0%.

TESTS

Assay (Total Nitrogen)   Proceed as directed under *Nitrogen Determination,* page 521.
α-Amino Nitrogen   Proceed as directed in the test for *α-Amino Nitrogen* under *Acid Hydrolyzed Proteins,* page 3, THIS SUPPLEMENT.
Arsenic   A *Sample Solution* prepared as directed for organic compounds meets the requirements of the *Arsenic Test,* page 464.
Aspartic and Glutamic Acids   Proceed as directed in the test for *Aspartic and Glutamic Acids* under *Acid Hydrolyzed Proteins,* page 3, THIS SUPPLEMENT.
Heavy Metals   Prepare and test a 1-g sample as directed in *Method II* under the *Heavy Metals Test,* page 513, using 20 μg of lead ion (Pb) in the control (*Solution A*).
Insoluble Matter   Transfer about 5 g, accurately weighed, into a 250-ml Erlenmeyer flask, add 75 ml of water, cover the flask with a watch glass, and boil gently for 2 min. Filter the solution through a tared filtering crucible, dry at 105° for 1 h, cool, and weigh.
Lead   A *Sample Solution* prepared as directed for organic compounds meets the requirements of the *Lead Limit Test,* page 518, using 10 μg of lead ion (Pb) in the control.
Sodium   Proceed as directed in the test for *Sodium* under *Acid Hydrolyzed Proteins,* page 3, THIS SUPPLEMENT.

Packaging and Storage   Store in well-closed containers.
Functional Use in Foods   Flavoring agent.

Bay Oil, page 33

Delete the *Requirement* and *Test* for *Solubility in Alcohol.*

Calcium Oxide, page 55

Change the *Requirement* for *Fluoride* to read:

Fluoride   Not more than 0.015%.

Carbon, Activated, page 70

Delete *Activated Charcoal* as a synonym.

Carmine, page 72

Replace the second sentence of the description with the following:

Cochineal consists of the dried female insects *Dactylopius coccus costa* (*Coccus cacti* L.) enclosing young larvae; the coloring principles derived therefrom consist chiefly of carminic acid ($C_{22}H_{20}O_{13}$).

Celery Seed Oil, page 78

Change the *Requirement* entitled *Acid Value* to read:

Acid Value   Not more than 4.5.

Change the *Requirement* entitled *Saponification Value* to read:

Saponification Value   Between 25 and 65.

Change the *Requirement* entitled *Specific Gravity* to read:

Specific Gravity   Between 0.870 and 0.910.

Disodium Guanylate, page 105

Insert the following *Test* for *Loss on Drying*:

Loss on Drying, page 518   Dry at 120° for 4 h.

Enzyme Preparations, page 107

Under *Coliforms,* page 110, change Section 46.039 to Section 46.016.

Fructose, page 130

In the *Test* entitled *Assay*, change the factor for the 200-mm tube from 0.562 to 0.555 in the last sentence of the paragraph.

Insert the following new monograph to precede the monograph entitled *Guar Gum*, page 141:

Grape Skin Extract

Enocianina

DESCRIPTION

Grape skin extract is a red to purple powder or liquid concentrate prepared by aqueous extraction of grape marc remaining from the pressing of grapes to obtain juice. Extraction is effected with water containing sulfur dioxide. After concentration by vacuum evaporation, the sugar content is reduced by fermentation: further concentration removes most of the alcohol. The primary color components are anthocyanins such as the glucosides of malvidin, peonidin, petunidin, delphinidin, or cyanidin. Other components naturally present are sugars, tartrates, malates, tannins, and minerals. Alcohol or sulfur dioxide may be added. The powder may contain an added carrier such as maltodextrin, modified starch, or gum. In acid solution, grape skin extract is red; in neutral to alkaline solution, it is violet to blue.

REQUIREMENTS

Identification

Transfer 1 g of sample and 1 g of potassium metabisulfite to a 100-ml volumetric flask, dissolve in about 50 ml of pH 3.0 *Citrate-Citric Acid Buffer* (see *Assay* below), and dilute to volume with the same buffer. The red color due to anthocyanins is bleached.

Assay The color strength (*CS*) expressed as the absorbance of a 1% solution in a cell of 1-cm pathlength at pH 3.0 shall not be less than 90% of the color strength as represented.

Arsenic Not more than 3 ppm.

Lead Not more than 10 ppm.

Pesticides Pesticide levels shall conform with national regulations in the country of use.

Sulfur Dioxide Not more than 0.1%.

TESTS

Assay Transfer about 0.2 g of grape skin extract, accurately weighed, to a 100-ml volumetric flask, dissolve in about 25 ml of pH 3.0 *Citrate-Citric Acid Buffer*, and dilute to volume

with the same buffer. (Prepare the buffer by adding 0.1 *M* sodium citrate dropwise to 0.1 *M* citric acid until a pH of 3.0 is reached, as determined by a glass electrode.) Allow this solution to stand in the dark for one-half hour, then remove any undissolved material by filtration or centrifugation. Adjust the pH to 3.0, and determine the absorbance of the clarified solution at 525 nm in a cell with a 1-cm pathlength. The color strength expressed as the absorbance of a 1% solution in a cell of 1-cm pathlength is calculated as:

$$CS = \text{Absorbance at 525 nm/Sample Weight in g.}$$

Arsenic A *Sample Solution* prepared as directed for organic compounds meets the requirements of the *Arsenic Test*, page 464.

Lead A *Sample Solution* prepared as directed for organic compounds meets the requirements of the *Lead Limit Test*, page 518, using 10 μg of lead ion (Pb) in the control.

Sulfur Dioxide Determine as directed in the general method, page 546. Omit preparation of the slurry in 300 ml of recently boiled and cooled water. Transfer the sample directly to the flask, dilute to 400 ml with water, and proceed as directed.

Packaging and Storage Store liquid grape skin extract in high-density polyethylene containers at 4°–14°. Store powdered grape skin extract in fiber drums at room temperature.

Functional Use in Foods Color.

L-Isoleucine, page 154

Change the *Molecular Weight* from 131.18 to 131.17.

L-Leucine, page 171

Change the *Molecular Weight* from 131.18 to 131.17.

Magnesium Oxide, page 178

Change the *Requirement* entitled *Heavy Metals (as Pb)* to read:

Heavy Metals (as Pb) Not more than 0.004%.

Insert the following revised sections for *Magnesium Sulfate* immediately preceding and following the *Tests*, which are unchanged:

Magnesium Sulfate, page 183

Epsom Salt

$MgSO_4 \cdot xH_2O$ Mol wt (anhydrous) 120.36

DESCRIPTION

Magnesium sulfate is produced with one or seven molecules of water of hydration, or in a dried form containing the equivalent of about 2.3 waters of hydration. It is in the form of a colorless crystal or a granular crystalline powder. It is odorless. It is readily soluble in water, slowly soluble in glycerine, and sparingly soluble in alcohol.

REQUIREMENTS

Identification

A 1 in 20 solution gives positive tests for *Magnesium*, page 517, and for *Sulfate*, page 517.

Assay   Not less than 99.5% of $MgSO_4$ after ignition.
Arsenic (as As)   Not more than 3 ppm.
Heavy Metals (as Pb)   Not more than 10 ppm.
Loss on Ignition   Between 13% and 16% for the monohydrate. Between 22% and 28% for the dried form. Between 40% and 52% for the heptahydrate.
Selenium   Not more than 0.003%.

Packaging and Storage   Store in well-closed containers.
Labeling   Label magnesium sulfate to indicate whether it is the monohydrate, the dried form, or the heptahydrate.
Functional Use in Foods   Nutrient; dietary supplement.

Mandarin Oil, Coldpressed, page 185

Change the *Requirement* entitled *Specific Gravity* to read:

Specific Gravity   Between 0.846 and 0.852.

Peppermint Oil, page 219

In the *Test* entitled *Assay for Total Menthol*, change the formula for calculating the percentage of total menthol from

$$7.814A(0.0021E)/(B - 0.021A)$$

to

$$7.814A(1 - 0.0021E)/(B - 0.021A).$$

Petroleum Wax, Synthetic, page 222

In the paragraph entitled *Ultraviolet Absorbance*, page 223, change the CFR reference from 21 CFR 121.1156 to 21 CFR 172.886.

Potassium Alginate, page 239

Change the *Requirement* entitled *Assay* to read:

Assay   It yields not less than 16.5% and not more than 19.5% of carbon dioxide (CO_2), corresponding to between 89.2% and 105.5% of potassium alginate (equiv wt 238.00), calculated on the dried basis.

Change the *Requirement* entitled *Heavy Metals* to read:

Heavy Metals (as Pb)   Not more than 0.004%.

Insert the following new *Requirement* to precede the *Requirement* entitled *Loss on Drying*:

Lead   Not more than 10 ppm.

Insert the following *Test* entitled *Lead* immediately preceding *Loss on Drying*:

Lead   A *Sample Solution* prepared as directed for organic compounds meets the requirements of the *Lead Limit Test*, page 518, using 10 µg lead ion (Pb) in the control.

Potassium Bicarbonate, page 239

Replace the *Test* entitled *Assay* with the following:

Assay   Weigh accurately about 4 g of sample, and dissolve it in 100 ml of water. Add 2 drops of methyl red TS, and titrate with 1 N hydrochloric acid. Add the acid slowly, with constant stirring, until the solution becomes faintly pink. Heat the solution to boiling, cool, and continue the titration until the pink color no longer fades after boiling. Each ml of 1 N hydrochloric acid is equivalent to 100.1 mg of $KHCO_3$.

Potassium Carbonate, page 240

Replace the *Test* entitled *Assay* with the following:

Assay Weigh accurately in a stoppered weighing bottle about 1 g of the dried sample obtained as directed under *Loss on Drying*, and dissolve it in 50 ml of water. Add 2 drops of methyl red TS, and titrate with 1 *N* hydrochloric acid. Add the acid slowly, with constant stirring, until the solution becomes faintly pink. Heat the solution to boiling, cool, and continue the titration until the faint pink color no longer fades after boiling. Each ml of 1 *N* hydrochloric acid is equivalent to 69.10 mg of K_2CO_3.

Potassium Nitrate, page 247

Change the *Requirement* entitled *Identification* to read:

Identification

A 1 in 10 solution gives positive tests for *Potassium*, page 517, and *Nitrate*, page 517.

Potassium Sorbate, page 252

Change the structure shown from $CH_3CH=CHCH=COOK$ to $CH_3CH=CHCH=CHCOOK$.

L-Serine, page 270

Change the *Molecular Weight* from 105.10 to 105.09.

Silicon Dioxide, page 271

Delete the *Requirement* entitled *Insoluble Substances*, page 272.

Sodium Alginate, page 274

Replace the last sentence of the *Requirement* entitled *Assay* with the following:

Each ml of 0.25 *N* sodium hydroxide consumed in the assay is equivalent to 27.75 mg of sodium alginate (equiv wt 222.00), calculated on the dried basis.

Sodium Bicarbonate, page 278

Replace the *Test* entitled *Assay* with the following:

Assay Weigh accurately about 3 g of sample, previously dried over silica gel for 4 h, and dissolve it in 100 ml of water. Add 2 drops of methyl red TS, and titrate with 1 *N* hydrochloric acid. Add the acid slowly, with constant stirring, until the solution becomes faintly pink. Heat the solution to boiling, cool, and continue the titration until the faint pink color no longer fades after boiling. Each ml of 1 *N* hydrochloric acid is equivalent to 84.01 mg of $NaHCO_3$.

Sodium Carbonate, page 280

Replace the *Test* entitled *Assay* with the following:

Assay Weigh accurately about 2 g of the dried salt, obtained as directed under *Loss on Drying*, and dissolve it in 50 ml of water. Add 2 drops of methyl red TS, and titrate with 1 *N* hydrochloric acid. Add the acid slowly, with constant stirring, until the solution becomes faintly pink. Heat the solution to boiling, cool, and continue the titration until the faint pink color no longer fades after boiling. Each ml of 1 *N* hydrochloric acid is equivalent to 53.00 mg of Na_2CO_3.

Sodium Saccharin, page 297

Replace the molecular structure with the following:

Spice Oleoresins, page 310

Insert the following new *Spice Oleoresins* in the proper alphabetical order:

Oleoresin Angelica Seed Obtained by the solvent extraction of the dried seed of *Angelica archangelica* L. as a dark brown or green liquid.

Oleoresin Anise   Obtained by the solvent extraction of the dried ripe fruit of *Pimpinella anisum* L. or *Illicium verum* L. as a dark brown or green liquid.

Oleoresin Basil   Obtained by the solvent extraction of the dried plant of *Ocimum basilicum* L. as a dark brown or green semisolid.

Oleoresin Caraway   Obtained by the solvent extraction of the dried seeds of *Carum carvi* L. as a green yellow to brown liquid.

Oleoresin Cardamom   Obtained by the solvent extraction of the dried seeds of *Elettaria cardamomum Maton* as a dark brown or green liquid.

Oleoresin Coriander   Obtained by the solvent extraction of the dried seeds of *Coriandrum sativum* L. as a brown yellow to green liquid.

Oleoresin Cubeb   Obtained by the solvent extraction of the dried fruit of *Piper cubeba* L. as a green or green brown liquid.

Oleoresin Cumin   Obtained by the solvent extraction of the dried seeds of *Cuminum cyminum* L. as a brown to yellow green liquid.

Oleoresin Dillseed   Obtained by the solvent extraction of the dried seeds of *Anethum graveolens* L. as a brown or green liquid.

Oleoresin Fennel   Obtained by the solvent extraction of the dried fruit of *Foeniculum vulgare Miller* as a brown green liquid.

Oleoresin Laurel Leaf   Obtained by the solvent extraction of the dried leaves of *Laurus nobilis* L. as a dark brown or green semisolid.

Oleoresin Marjoram   Obtained by the solvent extraction of the dried herb of the marjoram shrub *Majorama hortensis Moench* as a dark green to brown viscous liquid or semisolid.

Oleoresin Origanum   Obtained by the solvent extraction of the dried flowering herb *Origanum vulgare* L. as a dark brown green semisolid.

Oleoresin Parsley Leaf   Obtained by the solvent extraction of the dried herb of *Petroselinum crispum* L. as a brown to green liquid.

Oleoresin Parsley Seed   Obtained by the solvent extraction of the dried seeds of *Petroselinum crispum* L. as a deep green semiviscous liquid.

Oleoresin Pimenta Berries   Obtained by the solvent extraction of the dried fruit of *Pimenta officinalis Lindley* as a brown green to dark green liquid.

Oleoresin Thyme   Obtained by the solvent extraction of the dried flowering plant *Thymus vulgaris* L. as a dark brown to green, viscous semisolid.

Insert the following in the proper alphabetical order under *Additional Requirements*, page 311:

Oleoresin Angelica Seed   *Volatile Oil Content:* between 2 ml and 7 ml per 100 g.

Oleoresin Anise   *Volatile Oil Content:* between 9 ml and 22 ml per 100 g.

Oleoresin Basil   *Volatile Oil Content:* between 4 ml and 17 ml per 100 g.

Oleoresin Caraway   *Volatile Oil Content:* between 10 ml and 20 ml per 100 g.

Oleoresin Cardamom   *Volatile Oil Content:* between 50 ml and 80 ml per 100 g.

Oleoresin Coriander   *Volatile Oil Content:* between 2 ml and 12 ml per 100 g.

Oleoresin Cubeb   *Volatile Oil Content:* between 50 ml and 80 ml per 100 g.

Oleoresin Cumin   *Volatile Oil Content:* between 10 ml and 30 ml per 100 g.

Oleoresin Dillseed   *Volatile Oil Content:* between 10 ml and 20 ml per 100 g.

Oleoresin Fennel   *Volatile Oil Content:* between 3 ml and 20 ml per 100 g.

Oleoresin Laurel Leaf   *Volatile Oil Content:* between 5 ml and 25 ml per 100 g.

Oleoresin Marjoram   *Volatile Oil Content:* between 10 ml and 20 ml per 100 g.

Oleoresin Origanum   *Volatile Oil Content:* between 20 ml and 45 ml per 100 g.

Oleoresin Parsley Leaf   *Volatile Oil Content:* between 2 ml and 10 ml per 100 g.

Oleoresin Parsley Seed   *Volatile Oil Content:* between 2 ml and 7 ml per 100 g.

Oleoresin Pimenta Berries   *Volatile Oil Content:* between 20 ml and 50 ml per 100 g.

Oleoresin Thyme   *Volatile Oil Content:* between 5 ml and 12 ml per 100 g.

Spike Lavender Oil, page 311

Change the *Requirement* entitled *Esters* to read:

Esters   Not less than 1.5% and not more than 4.0% of esters calculated as linalyl acetate ($C_{12}H_{20}O_2$).

d-α-Tocopheryl Acetate Concentrate, page 335

Change the *Requirement* entitled *Identification* to read:

Identification

It meets the requirements of *Identification Tests A and B* under *d*-α-*Tocopheryl Acetate*, page 333.

Change the *Test* entitled *Assay* to read:

Assay Proceed as directed in the *Assay* under *d-α-Tocopheryl Acetate*, page 333, using the following as the *Assay Preparation*: Dissolve an accurately weighed amount of the sample equivalent to about 30 mg of *d-α*-tocopheryl acetate in 10.0 ml of the *Internal Standard Solution*.

Triacetin, page 337

Replace the sentence describing *Packaging and Storage*, page 338, with the following:

Packaging and Storage Store in well-closed containers.

Triethyl Citrate, page 339

In the *Test* entitled *Assay*, replace phenolphthalein with bromothymol blue in the fourth line from the bottom of the paragraph.

Xylitol, page 349

In the *Test* entitled *Other Polyols*, in the paragraph labeled *Procedure*, change the approximate retention time for mannitol hexaacetate from 30 min to 20 min.

Zinc Sulfate, page 351

Change the *Requirement* entitled *Assay*, for the monohydrate only, to read:

Assay *Monohydrate:* not less than 98.0% and not more than 100.5% of $ZnSO_4 \cdot H_2O$.

Change the last sentence of the *Test* entitled *Assay* to read:

Each ml of 0.05 *M* disodium EDTA is equivalent to 8.973 mg of $ZnSO_4 \cdot H_2O$ or 14.38 mg of $ZnSO_4 \cdot 7H_2O$.

3/ *Specifications for Flavor Aromatic Chemicals and Isolates*

Tabular Section, pages 354–419

Transfer the column entitled *Solubility in Alcohol* from *Requirements* to *General Information and Description*.

3-Acetyl-2,5-dimethyl Furan, page 354

(2,5-Dimethyl-3-acetylfuran)

[FEMA No. 3391]

Change the *Assay Min, %,* page 355, from 99.8% to 99.0%.

Allyl Hexanoate, page 354

(Allyl Caproate)

[FEMA No. 2032]

Change the *Solubility in Alcohol*, page 355, from 1 ml in 3.5 ml 70% alc to 1 ml in 6 ml 70% alc.

Allyl α-Ionone, page 356

(Allyl Ionone)

[FEMA No. 2033]

Change the *Solubility in Alcohol*, page 357, from 1 ml in 8 ml 70% alc gives clear soln to 1 ml in 1 ml 90% alc gives clear soln.

α-Amylcinnamaldehyde, page 356

(Amylcinnamaldehyde)

[FEMA No. 2061]

Change the *Solubility in Alcohol*, page 357, from 1 ml in 4.5 ml 80% alc to 1 ml in 5 ml 80% alc.

Anisyl Acetate, page 356

(*p*-Methoxybenzyl Acetate)

[FEMA No. 2098]

Change the *Sp. Gr.*, page 357, from 1.104–1.107 to 1.104–1.111.

Benzaldehyde, page 358

[FEMA No. 2127]

Change the formula in the column entitled *Assay Min, %*, page 359, from C_7H_8O to C_7H_6O.

Change *Solubility in Sulfite* under *Other Requirements*, page 359, to *Solubility in Bisulfite*.

Benzyl Acetate, page 358

[FEMA No. 2135]

Insert in the column entitled *A.V. Max*, page 359, (phenol red TS).

Benzyl Cinnamate, page 358

[FEMA No. 2142]

Change the *Assay Min, %*, page 359, from 99.0% of $C_{16}H_{14}O_2$ (M-6) to 98.0% of $C_{16}H_{14}O_2$ (M-6).

Benzyl Salicylate, page 360

[FEMA No. 2151]

Change the assay method, page 361, from (M-6) to (M-7).

Insert in the column entitled *A.V. Max*, page 361, (phenol red TS).

Change the *Solubility in Alcohol*, page 361, from 1 ml in 9 ml 90% alc to 1 ml in 5 ml 95% alc.

Butyl Acetate, page 360

(*n*-Butyl Acetate)

[FEMA No. 2174]

Change the *Ref. Index*, page 361, from 1.393–1.395 to 1.393–1.396.

Butyraldehyde, page 362

(Butyl Aldehyde)

[FEMA No. 2219]

Change the formula in the column entitled *Assay Min, %*, page 363, from C_6H_8O to C_4H_8O.

Camphene, page 362

[FEMA No. 2229]

Delete the requirements for *Ref. Index* and *Sp. Gr.*

Change the *Solidification Pt.* under *Other Requirements*, page 363, to 40° (p. 538).

Carvacrol, page 362

[FEMA No. 2245]

Change the *Sp. Gr.*, page 363, from 0.974–0.979 to 0.974–0.980.

β-Caryophyllene, page 362

[FEMA No. 2252]

Change the *Solubility in Alcohol*, page 363, from 1 ml in 4 ml 95% alc gives clear soln to 1 ml in 6 ml 95% alc gives clear soln.

Under *Other Requirements*, page 363, change *Phenols*—3% (M-33b) to *Phenols*—3% (M-10).

Cinnamaldehyde, page 364

(Cinnamic Aldehyde; Cinnamal)

[FEMA No. 2286]

Change the *A.V. Max*, page 365, from 5.0 to 10.0.

Cinnamyl Acetate, page 364

[FEMA No. 2293]

Change the *A. V. Max*, page 365, from 3.0 to 1.0.

Cinnamyl Formate, page 364

[FEMA No. 2299]

Change the *Solubility in Alcohol*, page 365, from 1 ml in 10 ml 70% alc, and in 2 ml 80% alc, gives clear soln to 1 ml in 2 ml 80% alc gives clear soln.

Citronellal, page 366

(3,7-Dimethyl-6-octen-1-al)

[FEMA No. 2307]

Under *Mol Wt/Formula/Structure*, page 366, change the structure to:

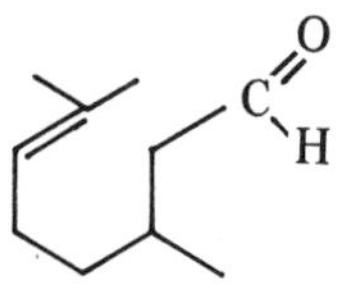

Citronellol, page 366

(3,7-Dimethyl-6-octen-1-ol)

[FEMA No. 2309]

Under *Other Requirements*, page 367, delete *Angular Rotation*—between −1° and +5° (p. 530, 100-mm tube).

Citronellyl Acetate, page 366

(3,7-Dimethyl-6-octen-1-yl Acetate)

[FEMA No. 2311]

Change the formula in the column entitled *Mol Wt/Formula/Structure*, page 366, from $C_{22}H_{22}O_2$ to $C_{12}H_{22}O_2$.

Citronellyl Formate, page 366

(3,7-Dimethyl-6-octen-1-yl Formate)

[FEMA No. 2314]

Change the *Ref. Index*, page 367, from 1.443–1.450 to 1.443–1.452.

Cuminic Aldehyde, page 366

(*p*-Cuminic Aldehyde; Cumaldehyde; *p*-Isopropylbenzaldehyde; Cuminal)

[FEMA No. 2957]

Remove the word "provided" from the column entitled *GLC Profile*.

Cyclamen Aldehyde, page 368

[2-Methyl-3-(*p*-isopropylphenyl)-propionaldehyde]

[FEMA No. 2743]

Change the *Solubility in Alcohol*, page 369, from 1 ml in 1 ml 80% alc to 1 ml in 3 ml 80% alc.

Δ-Decalactone, page 368

[FEMA No. 2361]

Change the assay method, page 369, from (M-8c) to (M-8a).

1-Decanol, Natural, page 368

(Decyl Alcohol; Alcohol C-10)

[FEMA No. 2365]

Change the name to *1-Decanol*.

Diethyl Succinate, page 370

[FEMA No. 2377]

Change the assay method, page 371, from (M-8c) to (M-8a).

Dimethyl Benzyl Carbinyl Butyrate, page 370

(α,α-Dimethylphenethyl Butyrate)

[FEMA No. 2394]

Change the assay method, page 371, from (M-8a) to (M-6).

Δ-Dodecalactone, page 372

[FEMA No. 2401]

Change the assay method, page 373, from (M-8c) to (M-8a).

Estragole, page 372

(*p*-Allylanisole)

[FEMA No. 2411]

Insert under *Name of Substance (Synonyms)*, page 372, Methylchavicol, immediately above [FEMA No. 2411].

Ethyl Cinnamate, page 374

(Ethyl 3-Phenylpropenate)

[FEMA No. 2430]

Change the *Assay Min, %*, page 375, from 99.0% of $C_{12}H_{12}O_2$ (M-6) to 98.0% of $C_{11}H_{12}O_2$ (M-6).

Change the *Ref. Index*, page 375, from 1.558–1.560 to 1.558–1.561.

Ethyl 2-Methylbutyrate, page 376

[FEMA No. 2443]

Change the assay method, page 377, from (M-8a) to (M-6).

Change the *Ref. Index*, page 377, from 1.396–1.400 to 1.393–1.400.

Change the *Sp. Gr.*, page 377, from 0.861–0.866 to 0.863–0.870.

Eugenol, page 380

(4-Allyl-2-methoxyphenol; Eugenic Acid; 4-Allylguaiacol)

[FEMA No. 2467]

Change the *Assay Min, %*, page 381, from 100% of phenols by vol (M-10) to 98% of phenols (M-8b).

Under *Other Requirements*, page 381, delete *Dist. Range*—NLT 95% between 250° and 255° (p. 478).

Eugenyl Acetate, page 380

(4-Allyl-2-methoxyphenyl Acetate; Eugenol Acetate; Acetyl Eugenol; Aceteugenol)

[FEMA No. 2469]

Remove the word "provided" from the column entitled *GLC Profile*.

Insert in the column entitled *A.V. Max*, page 381, (phenol red TS). See also the note in M-7, page 23, THIS SUPPLEMENT.

Farnesol, page 380

(3,7,11-Trimethyl-2,6,10-dodecatrien-1-ol)

[FEMA No. 2478]

Change the *Assay Min, %*, page 381, from 97% of $C_{15}H_{26}O$ (M-8a) to NLT 96% (sum of 4 major isomers) (M-8a).

Change the *Ref. Index*, page 381, from 1.487–1.489 to 1.487–1.492.

Change the *Sp. Gr.*, page 381, from 0.887–0.889 to 0.884–0.889.

Geranyl Acetate, page 380

(3,7-Dimethyl-2,6-octadien-1-yl Acetate)

[FEMA No. 2509]

Change the *Solubility in Alcohol*, page 381, from 1 ml in 8 ml 70% alc to 1 ml in 9 ml 70% alc.

Geranyl Benzoate, page 380

(3,7-Dimethyl-2,6-octadien-1-yl Benzoate)

[FEMA No. 2511]

Change the *Ref. Index*, page 381, from 1.513–1.581 to 1.513–1.518.

Geranyl Butyrate, page 382

(3,7-Dimethyl-2,6-octadien-1-yl Butyrate)

[FEMA No. 2512]

Change the *Solubility in Alcohol*, page 383, from 1 ml in 4 ml 80% alc to 1 ml in 6 ml 80% alc.

Geranyl Formate, page 382

(3,7-Dimethyl-2,6-octadien-1-yl Formate)

[FEMA No. 2514]

Change the method under *A.V. Max*, page 383, from M-15d to *General Method*, p. 499.

Geranyl Phenylacetate, page 382

(3,7-Dimethyl-2,6-octadien-1-yl Phenylacetate)

[FEMA No. 2516]

Change the *Ref. Index*, page 383, from 1.507–1.511 to 1.506–1.511.

Heptanal, page 382

(Aldehyde C-7; Heptaldehyde)

[FEMA No. 2540]

Change the *Assay Min, %*, page 383, from 90.0% of $C_7H_{14}O$ (M-2a) to 92.0% of $C_7H_{14}O$ (M-2a).

2-Heptanone, page 382

(Methyl Amyl Ketone)

[FEMA No. 2544]

Change the *Sp. Gr.*, page 383, from 0.813–0.818 to 0.814–0.819.

3-Heptanone, page 382

(Ethyl Butyl Ketone)

[FEMA No. 2545]

Under *Other Requirements*, page 383, change *Acidity*—0.02% (M-15a) to *Acidity*—0.1% (M-15a).

Hexanal, page 384

(Caproic Aldehyde; Hexaldehyde)

[FEMA No. 2557]

Change the *Ref. Index*, page 385, from 1.403–1.407 to 1.402–1.407.

Change the *Sp. Gr.*, page 385, from 0.808–0.812 to 0.808–0.817.

cis-3-Hexen-1-ol, page 384

[FEMA No. 2563]

Change the *Assay Min, %*, page 385, from 98.0% of $C_6H_{12}O$ (M-8a) to 98.0% as sum of isomers, 92.0% major component.

Hexyl Alcohol, Natural, page 386

(1-Hexanol; Alcohol C-6)

[FEMA No. 2567]

Change the name to *Hexyl Alcohol.*

Hexyl Isovalerate, page 386

[FEMA No. 3500]

Change the assay method, page 387, from (M-8a) to (M-6).

α-Ionone, page 388

[4(2,6,6-Trimethyl-2-cyclohexenyl)-3-butene-2-one]

[FEMA No. 2594]

Change the *Assay Min, %,* page 389, from 99.0% of $C_{13}H_{20}O$ (M-3) to 98.0% of $C_{13}H_{20}O$ (M-3).

β-Ionone, page 388

[4(2,6,6-Trimethyl-1-cyclohexenyl)-3-butene-2-one]

[FEMA No. 2595]

Change the *Assay Min, %,* page 389, from 90.0% of $C_{13}H_{20}O$ (M-3) to 97.0% of $C_{13}H_{20}O$ (M-3).

Isoamyl Butyrate, page 388

(Amyl Butyrate)

[FEMA No. 2060]

Change the *Sp. Gr.,* page 389, from 0.860–0.864 to 0.860–0.865.

Isoamyl Salicylate, page 388

(Amyl Salicylate)

[FEMA No. 2084]

Insert in the column entitled *A.V. Max,* page 389, (phenol red TS). See also the note in M-6, page 23, THIS SUPPLEMENT.

Isobutyl Acetate, page 390

[FEMA No. 2175]

Remove the word "provided" from the column entitled *GLC Profile.*

Isobutyl Alcohol, page 390

[FEMA No. 2179]

Insert in the column entitled *Assay Min, %,* page 391, 98.0% of $C_4H_{10}O$ (M-8a).

Isobutyric Acid, page 390

(2-Methyl Propanoic Acid; Isophenylformic Acid)

[FEMA No. 2222]

Under *Name of Substance (Synonyms)* change *Isophenylformic Acid* to *Isopropylformic Acid.*

Lauryl Alcohol, Natural, page 392

(1-Dodecanol; Alcohol C-12)

[FEMA No. 2617]

Change the name to *Lauryl Alcohol.*

Linalyl Propionate, page 394

(3,7-Dimethyl-2,6-octadien-3-yl Propionate)

[FEMA No. 2645]

Change the *Ref. Index*, page 395, from 1.450–1.455 to 1.449–1.454.

Change the *Solubility in Alcohol*, page 395, from 1 ml in 7 ml 70% alc to 1 ml in 2 ml 80% alc.

Change the *Sp. Gr.*, page 395, from 0.895–0.902 to 0.893–0.902.

Menthol, page 394

(3-*p*-Menthanol)

[FEMA No. 2665]

Under *Other Requirements*, page 395, insert the second *Solidification Pt.* for (*dl*-menthol) to read (*dl*-menthol)—27° to 28°, 30.5° to 32° (M-14b).

l-Menthone, page 396

(*l*-*p*-Menthan-3-one)

[FEMA No. 2667]

Change the assay method, page 397, from (M-8a) to (M-3).

dl-Menthyl Acetate, page 396

(*dl*-*p*-Menthan-3-yl Acetate)

[FEMA No. 2668]

Change the assay method, page 397, from (M-8a) to (M-6).

Change the *Ref. Index*, page 397, from 1.443–1.447 to 1.443–1.450.

l-Menthyl Acetate, page 396

(*l*-*p*-Menthan-3-yl Acetate)

[FEMA No. 2668]

Change the assay method, page 397, from (M-8a) to (M-6).

p-Methoxybenzaldehyde, page 396

(Anisic Aldehyde; *p*-Anisaldehyde)

[FEMA No. 2670]

Change the formulas, pages 396 and 397, from C_8H_8O to $C_8H_8O_2$.

Change the *Solubility in Alcohol*, page 397, from 1 ml in 7 ml 50% alc gives clear soln to 1 ml in 3 ml 60% alc gives clear soln.

2-Methoxypyrazine, page 396

[FEMA No. 3302]

Under *Other Requirements*, page 397, delete, *Dist. Range*—145°–150° (p. 478).

4′-Methyl Acetophenone, page 396

(Methyl *p*-Tolyl Ketone)

[FEMA No. 2677]

Change the *Assay Min, %*, page 397, from 98.0% of $C_9H_{10}O$ (M-3) to 95.0% of $C_9H_{10}O$ (M-3).

Change the *Ref. Index*, page 397, from 1.532–1.535 to 1.530–1.535.

Change the *Sp. Gr.*, page 397, from 1.001–1.004 to 0.996–1.004.

Methylbenzyl Acetate, page 398

(Tolyl Acetate So-Called)

Under *Mol Wt/Formula/Structure*, page 398, change the structure to:

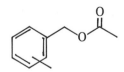

2-Methylbutyl Isovalerate, page 398

(2-Methylbutyl-3-methylbutanoate)

[FEMA No. 2753]

Change the *Assay Min, %*, page 399, from 94.0% of $C_{10}H_{20}O_2$ (M-8a) to 98% of $C_{10}H_{20}O_2$ (M-6).

α-Methylcinnamaldehyde, page 398

[FEMA No. 2697]

Change the *Solubility in Alcohol*, page 399, from 1 ml in 2 ml 70% alc remains clear on dilution to 1 ml in 3 ml 70% alc remains clear on dilution.

Methyl Eugenol, page 400

(Eugenyl Methyl Ether; 1,2-Dimethoxy-4-allylbenzene)

[FEMA No. 2475]

Insert in the column entitled *Assay Min, %*, page 401, 98% of $C_{11}H_{14}O_2$ (M8b).

Under *Other Requirements*, page 401 change *Eugenol*—1.0% (M-33a) to *Eugenol*—1% (M-8b).

Methyl 2-Methylbutyrate, page 400

(Methyl 2-Methylbutanoate)

[FEMA No. 2719]

Change the assay method, page 401, from (M-8a) to (M-6).

Methyl Salicylate, page 402

[FEMA No. 2745]

Insert in the column entitled *A.V. Max*, page 403, (phenol red TS). See also the note in M-6, page 23, THIS SUPPLEMENT.

1-Octanol, Natural, page 406

(Alcohol C-8; Octyl Alcohol; Capryl Alcohol)

[FEMA No. 2800]

Change the name to *1-Octanol*.

1-Octen-3-yl Acetate, page 406

[FEMA No. 3582]

Change the assay method, page 407, from (M-8a) to (M-6).

1-Octen-3-yl Butyrate, page 406

[FEMA No. 3612]

Change the assay method, page 407, from (M-8a) to (M-6).

3-Octyl Acetate, page 408

[FEMA No. 3583]

Change the assay method, page 409, from (M-8a) to (M-6).

Octyl Formate, page 408

[FEMA No. 2809]

Change the *Sp. Gr.*, page 409, from 0.869–0.872 to 0.869–0.874.

Phenethyl Isovalerate, page 408

[FEMA No. 2871]

Change the *Solubility in Alcohol*, page 409, from 1 ml in 11 ml 70% alc gives clear soln to 1 ml in 3 ml 80% alc gives clear soln.

2-Phenethyl 2-Methylbutyrate, page 408

[FEMA No. 3632]

Change the assay method, page 409, from (M-8a) to (M-6).

Phenoxyethyl Isobutyrate, page 410

[FEMA No. 2873]

Change the *Solubility in Alcohol*, page 411, from 1 ml in 2 ml 70% alc gives clear soln to 1 ml in 3 ml 70% alc gives clear soln.

Phenylacetaldehyde, page 410

(α-Toluic Aldehyde)

[FEMA No. 2874]

Change the *Ref. Index*, page 411, from 1.524–1.532 to 1.525–1.545.

Change the *Sp. Gr.*, page 411, from 1.025–1.035 to 1.025–1.045.

Rhodinyl Acetate, page 414

[FEMA No. 2981]

Change the *Ref. Index*, page 415, from 1.453–1.458 to 1.450–1.458.

Terpineol, page 416

(Menthen-1-ol-8)

[FEMA No. 3045]

Change the *Assay Min, %*, page 417, from 96.0% of $C_{10}H_{18}O$ as mixed isomers, principally alpha (M-8a), to 96.0% of $C_{10}H_{18}O$ as sum of isomers (M-8a).

Under *Other Requirements*, page 417, delete *Dist. Range*—NLT 90% within a 5° range between 214° and 224° (p. 478); and *Solidification Pt.*—NLT 2° (p. 538).

Tetrahydrolinalool, page 416

(3,7-Dimethyl-3-octanol)

[FEMA No. 3060]

Change the *Sp. Gr.*, page 417, from 0.923–0.927 to 0.823–0.829.

γ-Valerolactone, page 418

[FEMA No. 3103]

Change the assay method, page 419, from (M-8a) to (M-6).

Change the *Sp. Gr.*, page 419, from 1.045–1.050 to 1.047–1.054.

New flavor aromatic chemicals and isolates follow on next page.

General Information and Description

Name of Substance (Synonyms)	Mol Wt/Formula/ Structure	Physical Form/Odor	Solubility/B.P.	GLC Profile	Solubility in Alcohol
Allyl Heptanoate (Allyl Heptoate) [FEMA No. 2031]	$170.25/C_{10}H_{18}O_2/$ $CH_3(CH_2)_5COOC_3H_5$	colorless to pale yellow liq/fruity, sweet, pineapple			
Benzodihydropyrone [FEMA No. 2381]	$148.16/C_9H_8O_2/$	colorless to pale yellow liq/coconutlike			
Butyl Isovalerate [FEMA No. 2218]	$158.24/C_9H_{18}O_2/$ $(CH_3)_2CHCH_2COOC_4H_9$	colorless to pale yellow liq/fruity			
Dibenzyl Ether [FEMA No. 2371]	$198.25/C_{14}H_{14}O_2/$	colorless to pale yellow liq/earthy			
Ethyl Isobutyrate [FEMA No. 2428]	$116.16/C_6H_{12}O_2/$ $(CH_3)_2CHCOOC_2H_5$	colorless liq/fruity			
Ethyl Myristate [FEMA No. 2445]	$256.42/C_{16}H_{32}O_2/$ $CH_3(CH_2)_{12}COOC_2H_5$	colorless to pale yellow liq/waxy			
n-**Hexyl Acetate** [FEMA No. 2565]	$144.21/C_8H_{16}O_2/$ $CH_3(CH_2)_5OOCCH_3$	colorless liq/fruity			
4-(*p*-Hydroxyphenyl)-2-butanone [FEMA No. 2588]	$164.20/C_{10}H_{12}O_2/$	white solid/raspberry			

Requirements

I.D. Test	Assay Min, %	A.V. Max	Ref. Index	Sp. Gr.	Other Requirements
	97% (M-6)	1.0	1.426–1.430	0.880–0.885	**Allyl Alcohol**–NMT 0.1% by GLC.
			1.555–1.559	1.186–1.192	**Congealing Pt.**–NLT 22°C.
	97% (M-6)	1.0	1.407–1.411	0.856–0.859	
	98% (M-8b)		1.557–1.565	1.039–1.044	**Chlorinated Cmpd.**–passes test (p. 500).
	98% (M-6)	1.0	1.385–1.391	0.862–0.868	
	98% (M-6)	1.0	1.434–1.438	0.857–0.862	
	98% (M-6)	1.0	1.407–1.411	0.868–0.872	
	98% (M-3)				**Melting Range**–81°–86°C.

4/ Test Methods for Flavor Aromatic Chemicals and Isolates

M-3 Assay by Determination of Aldehydes and Ketones—Hydroxylamine Method, page 422

Insert in the appropriate alphabetical position:

l-Menthone: 1.0 g/77.12

M-6 Assay by Determination of Esters, page 423

Insert each of the following substances in its appropriate alphabetical position:

Allyl Heptanoate: 1.3 g/85.1
Butyl Isovalerate: 1.2 g/79.12
Dimethyl Benzyl Carbinyl Butyrate: 1.7 g/110.15
Ethyl Isobutyrate: 1.0 g/58.08
Ethyl 2-Methylbutyrate: 0.9 g/65.10
Ethyl Myristate: 1.9 g/128.21
n-Hexyl Acetate: 1.1 g/72.1
Hexyl Isovalerate: 1.4 g/93.15
Isoamyl Salicylate: 1.3 g/86.14 (NOTE: Use phenol red TS as the indicator.)
dl-Menthyl Acetate: 1.5 g/99.16
l-Menthyl Acetate: 1.5 g/99.16

2-Methylbutyl Isovalerate: 1.3 g/86.14
Methyl 2-Methylbutyrate: 0.9 g/58.08
Methyl Salicylate: 2 g/76.08 (NOTE: Use 50.0 ml of 0.5 N alcoholic potassium hydroxide, reflux for 2 h, and use phenol red TS as the indicator.)
1-Octen-3-yl Acetate: 1.3 g/85.12
1-Octen-3-yl Butyrate: 1.5 g/99.16
3-Octyl Acetate: 1.3 g/86.14
2-Phenethyl 2-Methylbutyrate: 1.6 g/103.14
γ-Valerolactone: 0.8 g/50.06

Remove the following substance from the list:

Benzyl Salicylate: 1.4 g/114.1 (NOTE: Reflux for 2 h, and use phenol red TS as indicator.)

M-7 Assay by Determination of Esters, High-Boiling Method, page 424

Insert each of the following substances in its appropriate alphabetical position:

Benzyl Salicylate: 1.4 g /114.1 (NOTE: Use phenol red TS as the indicator.)
Cresyl Acetate: 1.2g/75.09 (NOTE: Use phenol red TS as the indicator.)
Eugenyl Acetate: 1.7g/103.1 (NOTE: Use phenol red TS as the indicator.)

M-8 Assay by Gas-Liquid Chromatography, page 424

M-8c MISCELLANEOUS PROCEDURES

Remove the procedures for the following substances:

Δ-Decalactone, page 424
Diethyl Succinate, page 425
Δ-Dodecalactone, page 425

M-10 Assay for Determination of Phenols, page 425

Insert β-*Caryophyllene* immediately following *Carvacrol*.

Remove *Eugenol* from the list.

M-33 Limit Test for Phenolic Impurities, page 433

M-33a TEST FOR FREE PHENOLS

Remove *Methyl Eugenol* from the list.

M-33b TEST FOR PHENOLS USING CASSIA FLASK METHOD

Delete the test.

5/ *GLC Analysis of Flavor Aromatic Chemicals and Isolates*

Benzyl Butyrate, page 440

Change polar to nonpolar under the column headed *Column Type.*

Benzyl Propionate, page 440

Change polar to nonpolar under the column headed *Column Type.*

Isoamyl Formate, page 449

Delete the reference to component 7, *n*-Amyl Formate, under the column entitled *Component* for test EOA Type B.

6/ *General Tests and Apparatus*

Gas Chromatography, page 475

In the section entitled *Qualitative Analysis*, page 475, change the equation on page 476 from

$$R = 2(t_2 - t_1)/(w_1 - w_2)$$

to

$$R = 2(t_2 - t_1)/(w_1 + w_2).$$

Chromatography, page 471

Insert the following new subsection immediately following *High-Pressure Liquid Chromatography*, page 476, and immediately preceding the *1,4-Dioxane Limit Test*, page 477.

SYSTEM SUITABILITY TESTS FOR GAS AND HIGH-PERFORMANCE LIQUID CHROMATOGRAPHY

To determine if a chromatographic system is effective and gives reproducible results, it must be subjected to a test designed to judge its suitability at the time of use. The essence of such a test is the concept that the electronics, the equipment, the specimens, and the analytical operations all constitute a single analytical system that is amenable to a system suitability test. Specific data are collected from replicate injections of the assay preparation or standard preparation. These are used to calculate values of parameters that are compared with specified maximum and minimum values of parameters such as efficiency,

internal precision, resolution, retention time, nature of the calibration curve, response, and recovery, as specified in the individual monograph.

A useful parameter determined in the test is the reproducibility of replicate injections of the analytical reference solution, prepared from the reference standard as directed in the individual monograph and derivatized or mixed with the internal standard where applicable. The reproducibility of replicate injections is best expressed as the relative standard deviation. The calculation is expressed by the equation

$$S_R(\%) = \frac{100}{\overline{X}} \left[\frac{\sum\limits_{i=1}^{N} (X_i - \overline{X})^2}{N - 1} \right]^{\frac{1}{2}}$$

in which S_R is the relative standard deviation in percent, $\overline{X}$ is the mean of the set of N measurements, and X_i is an individual measurement. When an internal standard is used, the measurement X_i usually refers to the measurement of relative area, A_s,

$$X_i = A_s = a_r/a_i,$$

in which a_r is the area of the peak corresponding to the standard substance and a_i is the area of the peak corresponding to the internal standard. When peak heights are used, the measurement X_i refers to the measurement of relative heights, H_s,

$$X_i = H_s = h_r/h_i,$$

where h_r is the height of the peak corresponding to the standard substance and h_i is the height of the peak corresponding to the internal standard.

It is sometimes useful to specify a tailing factor to limit the maximum permissible asymmetry of the peak. Here the tailing factor, T, is defined as the ratio of the distance from the leading edge to the trailing edge of the peak ($W_{0.05}$) divided by twice the distance, f, measured from the position of the peak maximum to the leading edge of the peak, the distances being measured on a line drawn 5% of the peak height above the baseline (see Fig. 32, THIS SUPPLEMENT). For a symmetrical peak, the tailing factor is unity, whereas the value of T increases as asymmetry becomes more pronounced. The calculation is expressed by the equation

$$\text{tailing factor} = T = W_{0.05}/2f.$$

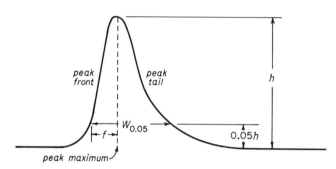

FIGURE 32   Asymmetrical Chromatographic Peak

Resolution factors, R, are specified to ensure separation of closely eluting components or to establish the separating efficiency of the system (see page 476).

Specification of columns and instrumental parameters in the individual monograph does not preclude the use of other suitable operating conditions. Normal variations in equipment and material may require adjustment of the experimental conditions to obtain acceptable operation.

7/ Solutions and Indicators

Test Solutions (TS) and Other Reagents, page 558

Insert the following test solution immediately preceding *Carr-Price Reagent*, page 559:

Calcium Sulfate TS A saturated solution of calcium sulfate in water.

8/ General Information

No Change.

9/ *Infrared Spectra*

Series C: Other Substances, page 713

Insert the spectrum of *Aspartame* immediately preceding
Butadiene-Styrene 75/25 Rubber, page 714:

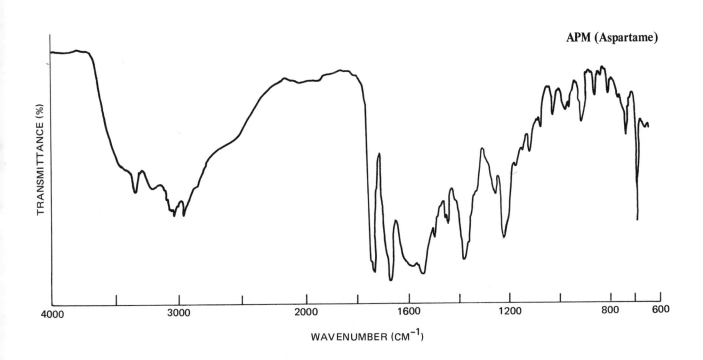

APM (Aspartame)

Insert the spectrum of *Thiamin Mononitrate* immediately
following *Rice Bran Wax*, page 721:

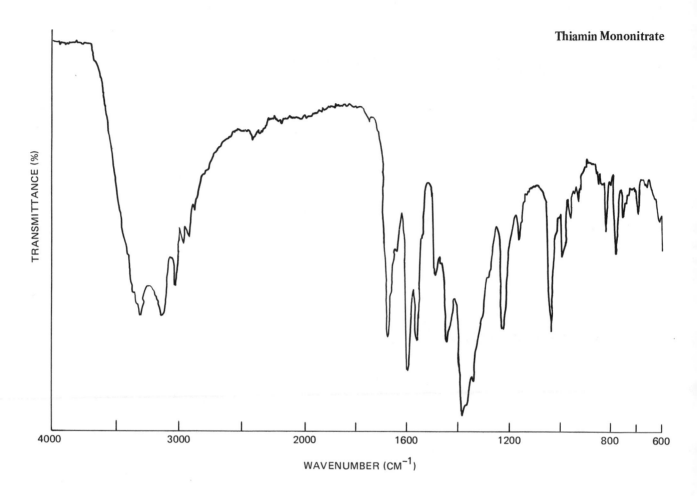

Thiamin Mononitrate

Index

An asterisk (*) indicates a new listing.